THE WORLD OF **STEM**

# The STEM of High-Speed Trains

By Derek Miller

Cavendish Square
New York

Published in 2021 by Cavendish Square Publishing, LLC
243 5th Avenue, Suite 136, New York, NY 10016



First Edition



Library of Congress Cataloging-in-Publication Data

Names: Miller, Derek, author.
Title: The STEM of high-speed trains / Derek Miller.
Description: New York : Cavendish Square, 2021. | Series: The world of STEM | Includes bibliographical references and index.
Identifiers: LCCN 2019042081 (print) | LCCN 2019042082 (ebook) | ISBN 9781502650221 (library binding) | ISBN 9781502650207 (paperback) | ISBN 9781502650214 (set) | ISBN 9781502650238 (ebook)
Subjects: LCSH: High speed trains–Juvenile literature.
Classification: LCC TF1455 .M55 2021 (print) | LCC TF1455 (ebook) | DDC 625.2–dc23
LC record available at https://lccn.loc.gov/2019042081
LC ebook record available at https://lccn.loc.gov/2019042082

Editor: Caitlyn Miller and Jennifer Lombardo
Copy Editor: Alex Tessman
Designer: Andrea Davison-Bartolotta

The photographs in this book are used by permission and through the courtesy of: Cover, p. 1 muratart/ Shutterstock.com; series art (background texture) bestbrk/Shutterstock.com; p. 4 Serjio74/Shutterstock.com; p. 6 (bottom) Country Gate Productions/Shutterstock.com; p. 6 (top) rawf8/Shutterstock.com; p. 7 CatwalkPhotos/Shutterstock.com; p. 9 (inset) Velocicaptor/Wikimedia Commons; p. 9 (main) © iStockphoto.com/Serjio74; p. 10 Igor Grochev/ Shutterstock.com; pp. 12–13 Anton Gvozdikov/Shutterstock.com; p. 15 Christopher May/Alamy Stock Photo; p. 16 TonyV3112/Shutterstock.com; p. 18 Blanscape/Shutterstock.com; p. 21 Ingus Kruklitis/Shutterstock.com; p. 22 Luka Hercigonja/Shutterstock.com; p. 23 © iStockphoto.com/ymgerman; p. 24 cyo bo/Shutterstock.com; p. 26 Studio BKK/ Shutterstock.com; p. 28 Kate Allen/Toronto Star via Getty Images; p. 29 Devrimb/iStock/Getty Images Plus/Getty Images.



CPSIA compliance information: Batch #CS20CSQ: For further information contact Cavendish Square Publishing LLC, New York, New York, at 1-877-980-4450.

Printed in the United States of America

# Contents

# AMAZING TRAINS

1

High-speed trains are a wonder of modern **engineering**. They can carry people and goods at speeds of more than 250 miles (400 kilometers) per hour! Around the world, high-speed trains connect people and businesses.

## *Moving Freight*

People need food to eat, clothes to wear, and **electronics** to use. These goods are called freight when they're shipped around the world. Freight is moved on trains, trucks, ships, and airplanes.

In the United States, more than 56 tons (51 metric tons) of freight are moved each year for every person. That much freight weighs

*Many people take trains to get places quickly.*

more than nine adult elephants! Trains carry a great deal of that freight.

Trains work by moving cars along a track. Today, electricity often powers high-speed trains. A long time ago, other power **sources** were used to move goods and people. More than 2,500 years ago, the first tracks were used to move goods. In Greece, grooves were cut into stone. Then, people and animals pulled carts along the grooves. The grooves worked like train tracks work today. They kept the carts steady.

Now, high-speed

*Freight trains carry goods across long distances.*

trains move people and goods. They're much faster than older trains, but they solve, or fix, the same problem—how to move things around the world.

## *The Problem of Friction*

Trains are huge and weigh a lot. This creates a problem. As the wheel of a train moves over a track, it causes friction. Friction happens when one thing rubs against another. Friction causes noise and heat. It also slows things down, which isn't good for high-speed trains. The heavier something is, the more friction it creates.

*Train wheels create noise and heat when they rub against the track.*

Since train cars are very heavy, they create a lot of friction. That's why trains are noisy. It's also why the wheels heat up. The friction wears down the metal train tracks. If the tracks get too worn down, they can't work anymore.

Engineers have come up with a clever way to fix this problem. Train tracks have greasers next to them. Greasers are machines that put grease on the tracks. They do this on their own. The grease is slippery. It makes it so there's less friction. The train makes less noise and goes faster when it moves over the grease. It also keeps the track from wearing down too much.

### THINK ON YOUR OWN

***Why do trains take so long to come to a stop after the train operator brakes?***

## THINK ON YOUR OWN

**Why do some trains spread sand on the tracks in front of their wheels when it rains?**

Greasers apply grease to the train tracks. The pipes shown here are part of this system.

# CHANGES OVER TIME

2

The first modern trains began running in the 1800s. These early trains were similar to trains today. They ran on metal tracks. However, they moved very slowly. Over time, new **materials** and methods made them faster.

## Powering Trains

In the 1800s, steam engines powered trains. Steam from boiling water provided the energy, or power, that moved locomotives. A locomotive is the train car that powers a train. Most train cars don't have any power themselves. They need the locomotive to push or pull them.

*Clouds of steam rise out of the top of a steam locomotive.*

Train tracks cross each other so trains can change direction.

Steam engines didn't have enough power to allow trains to move as fast as they do today. Now, electricity allows trains to go much faster. However, electricity wasn't all that was needed for high-speed trains. The tracks needed to be made better as well.

## *Track Materials*

Early railway tracks were wooden. They wore down very quickly from the weight of the train cars. This material wasn't very good at its job.

Soon, engineers began using iron tracks. The iron was much stronger than wood, but it still wasn't strong enough. Iron tracks couldn't **support** heavy, fast-moving trains.

In the 1860s, scientists figured out how to make steel cheaply. Steel is stronger than iron. This new material was key to the golden age of trains. Steel tracks were built across the United States. Today, tracks are still made out of steel.

## Track Technologies

**THINK ON YOUR OWN**

***Are there any other materials that might make good train tracks?***

High-speed trains ran into other problems with their tracks. Early tracks were made by laying rails in a line. The metal rails were pushed up against each other. The tiny gap between the rails wasn't a big problem. It created some noise when the train wheels went over it. The train car also shook a little bit. However, the trains were slow enough that this wasn't too uncomfortable.

With high-speed trains, the tiny gaps became a serious, or big, problem. The trains hit the gaps much faster. The train cars shook more and made loud noises. Engineers needed to fix this problem. They began welding the gaps. Welding is when someone uses

*Shown here is a track worker welding the track so there are no gaps between sections.*

heat to join two materials by melting them together. Welding the rail gaps led to a smoother, quieter ride.

*High-speed train tracks don't have sharp corners. Instead, they change direction gradually.*

The tracks also needed to be set up differently for high-speed trains. The faster a train is moving, the harder it is to turn. A slow train can go around a sharp corner. A fast train will derail—or go off the tracks—if it goes around the same corner. Train drivers slow down around bends.

High-speed trains need to go fast. They can't slow down often to make turns. That's why high-speed train tracks turn very gradually. The track takes a long time to form a small bend. This lets the train go fast the whole time.

## THINK ON YOUR OWN

***Why is it a problem when high-speed trains share railway tracks with slower trains?***

# CREATING HIGH-SPEED TRAINS

3

The first high-speed rail system was built in Japan. It's called the Shinkansen. Americans call this type of train a bullet train because it goes fast, like a bullet shot out of a gun. Opening in 1964, it carried passengers at speeds of up to 130 miles (210 km) per hour. The Shinkansen soon linked cities across Japan. Today, Shinkansen trains reach speeds of 200 miles (320 km) per hour! Many people like these trains because traveling now takes much less time.

## Trains and Cities

Trains are a very good way to move large numbers of people. They use less energy than

*Japan is home to the world's first high-speed, or bullet, train system.*

cars or airplanes. This is because many people can fit in a train car.

The downside of trains is that they must follow a track. That's why they're best used in cities and to connect cities that are close to one another. When cities are very spread out, the way many American cities are, trains are less useful. It's easier for people to drive their car if they live far from a city.

Passenger trains are a slower way to travel over a long distance than planes. It's not a big problem if freight takes days to arrive, but people want to go fast. Many people would rather fly across the United States in less than a day than take a train for many days. This is why trains aren't as useful in very large countries.

## THINK ON YOUR OWN

***Why do engineers want to use less energy to power trains?***

Japan is the perfect country for high-speed trains. Many large cities are close to one another. The Shinkansen links them together so people can take a high-speed train each day to work instead of driving.

## A New Design

Over time, Shinkansen trains got faster and faster. Engineers wanted to make it quicker for people to travel by train. New technologies made this possible. That speed came with a problem, though.

The Shinkansen trains go through many tunnels in cities. As they got faster, the trains became noisier.

*Trains go through tunnels so they don't need to go around water or mountains.*

When they went through the tunnels, the trains created a **sonic boom** just like an airplane! The air swirled around the trains and made the noise. People living near the tunnels weren't happy.

To fix the problem, engineers looked to nature. Eiji Nakatsu was a birdwatcher. He thought of a kingfisher—a bird that dives into water at high speed. Because of the shape of its beak, there's barely a splash when it hits the water.

*The beak of the kingfisher inspired the shape of the Shinkansen.*

Nakatsu **designed** a train that was shaped the same way as a kingfisher's beak. Other engineers tested his design. They found that it stopped the air from swirling in the tunnels and causing a sonic boom. The front of the trains was changed. The problem was solved!

*The design of this high-speed train makes it less noisy when it goes through tunnels.*

## THINK ON YOUR OWN

**Why do people in Japan use fast trains to travel between cities instead of airplanes?**

# TRAINS TODAY AND TOMORROW

4

There's a lot of interest in trains today. Scientists and governments are worried about **climate change**. They want people to use less energy, especially energy from **fossil fuels**. Trains use less energy than cars or airplanes, so they're often seen as better for the earth. Some newer trains don't use fossil fuels at all.

## *Maglev Trains*

Maglev technology is likely to change the way trains run around the world. "Maglev" stands for "magnetic levitation." "Levitation" is a word that means making something float. In a maglev train,

*It doesn't look like it, but this maglev train in China is actually floating above its track!*

magnets let the train float above its tracks! This means the train causes less friction and doesn't waste as much energy as other kinds of trains. Maglev trains also offer a smooth ride for passengers because of the reduced, or lower, friction.

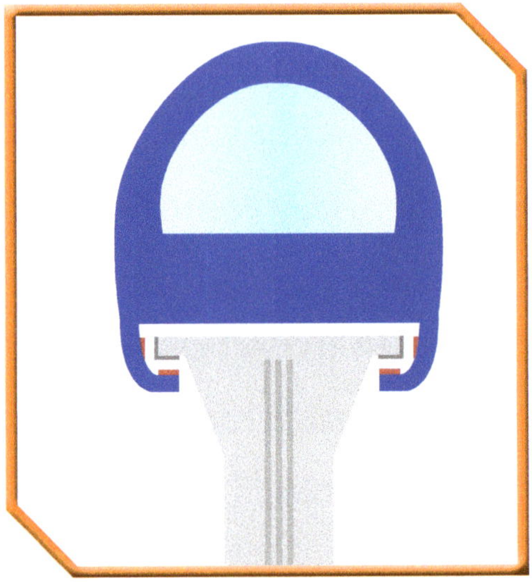

*This diagram shows how a maglev train levitates. Instead of touching the track (gray), the train (blue) floats just above it due to its magnets (red).*

There are already some maglev trains around the world. They're mostly in Asian countries such as Japan, South Korea, and China. As engineers work on this technology, they may soon be used in other places too.

China is working on a new maglev train. It hopes to begin building the train in 2021. This train will be able to travel at speeds of 370 miles (600 km) per hour!

China hopes the new maglev train will make train travel between major cities faster than air travel. It will also use less energy than air travel.

**THINK ON YOUR OWN**

***Why is the ride smoother in a maglev train than a normal train?***

## *The Hyperloop*

Another new technology that's being planned for the future is the Hyperloop. Like maglev trains, some plans for a Hyperloop system call for magnets, but instead of using a track, it will use a tube. Inside the tube, the air will be removed, or taken out. When there's no air somewhere, it's called a vacuum.

The vacuum makes it so there's less friction. We often think of friction as two objects rubbing against each other, but even air causes friction when an object moves. At very high speeds, the friction from air slows things down, creating a limit to how fast trains can go. The Hyperloop will solve this by using a vacuum tube. Pods with air inside will carry people and goods through the vacuum tube.

*In the future, people might move through a vacuum tube such as this one.*

Unlike maglev trains, the Hyperloop isn't working yet. Scientists are hopeful it will be built one day. If it is, it could move people at speeds of up to 670 miles (1,080 km) per hour. Maybe new kinds of trains will go even faster in the future!

*Shown here is a picture of what a Hyperloop pod might look like in the future.*

## THINK ON YOUR OWN

**Why can't people live in a vacuum?**

# Glossary

**climate change** Change in Earth's weather caused by human activity.

**design** The act of creating a plan for something. Also, the plan itself.

**electronics** Devices that operate through the use of many small electrical parts.

**engineering** The study and practice of using math and science to do useful things, such as building machines.

**fossil fuel** A fuel—such as coal, oil, or natural gas—that is formed in the earth from dead plants or animals.

**material** Something from which other things can be made.

**sonic boom** The loud bang that something breaking the speed of sound makes.

**source** A person, place, or thing from which something comes or where it can be found.

**support** To hold the weight of something.

# Find Out More

## Books

Bankston, John. *Japanese Bullet Train*. Hallandale, FL: Little Mitchie, 2020.

Ciovacco, Justine. *All About Trains*. New York, NY: Britannica, 2017.

## Websites

**History of Trains**

*dkfindout.com/us/transportation/history-trains*

Learn all about how trains have changed over time on this website by DK.

**Trains**

*online.kidsdiscover.com/unit/trains*

Explore interesting train facts and photos on this website presented by Kids Discover.

**Publisher's note to educators and parents:** Our editors have carefully reviewed these websites to ensure that they are suitable for students. Many websites change frequently, however, and we cannot guarantee that a site's future contents will continue to meet our high standards of quality and educational value. Be advised that students should be closely supervised whenever they access the Internet.

# Index